THE FUTURE OF
Transportation

By Lori Dittmer

CREATIVE EDUCATION

TABLE OF Contents

Introduction...4

Walking, Wheels, and World Travel...6

    The Compass Points the Way...13

    First in Space...14

Enhancing Safety and Efficiency...16

    Competition Fuels Improvement...23

    The Risks of Discovery...25

Better Fuels and Farther Flights...28

    Food Versus Fuel...30

    Global Positioning Progress...33

From Here to There...36

    A Glimpse of the Future...39

    Magellan Circles the World...42

Glossary...44

Selected Bibliography...46

Web Sites...46

Index...47

INTRODUCTION

When you are ready for school each morning, do you walk, pedal your bike, ride in a car, or take the bus? Imagine hopping into your own transporter, powered by electricity or perhaps solar energy, and quietly gliding down the street. If there were too many cars on the road, you could simply steer up into the air and fly the rest of the way. Other vehicles would be zooming around too, but everyone would be following the rules of air space. The Jetsons were a cartoon family that lived in space, flying their small, personal ships from place to place. Reality might someday catch up with this fantasy, especially if people set up **colonies** to live in space.

For now, you'll have to stick to traditional modes of transportation, such as cars and buses, as you go to school, the grocery store, and other destinations. But in the future, you might find yourself racing across the country on a high-speed train, traversing the ocean in a personal submarine, or rocketing to the moon. For thousands of years, the way people travel has improved by leaps and bounds. Instead of walking along paths beaten down by animals, people can drive at high speeds on smooth, paved highways. One hundred years ago, the very first cars and airplanes were being developed; today, scientists are eyeing trips to Mars. What does the future hold next?

Although people have long dreamed of the day when flying-saucer-type transportation will become reality, most transportation will remain on the ground for a while

WALKING, WHEELS, AND WORLD TRAVEL

Transportation refers to the movement of people or things from one place to another. To travel efficiently, people need two things: a vehicle for **propulsion**, such as a bicycle, horse, or car; and **infrastructure**, including roads, highways, and bridges on which to travel. Ancient people traveled by foot on dirt paths trampled by animals. They used elephants, camels, and other beasts of burden to pull cargo on sledges, two flat boards that slid along the path. The earliest known road is the Sweet Track, estimated to be 6,000 years old. Discovered accidentally in 1970 by an English peat cutter named Ray Sweet, the road was built from tree branches and planks made from tree trunks and likely was constructed to help people cross the wet lowlands of southwestern England.

The ancient Romans improved on road building when, in 312 B.C., ruler Appius Claudius Caecus ordered the construction of a vast highway network that would eventually include 370 major roads covering 50,000 miles (80,470 km). Romans often built their roads by layering packed earth, stone blocks, and sand. Some of the sturdiest were five feet (1.5 m) thick with slight arches at the middle so water would drain off the surface of the road. Today's roads are built with this same **cambered** design.

The Silk Road was the most famous of all ancient road systems. Used for nearly 10 centuries, the Silk Road ran through China and central Asia and also led to India, Saudi Arabia, and Egypt. Along the

Stone and wood formed some of the earliest travel routes, including ancient Roman roadways (left) and the Sweet Track (pictured in reconstructed form, below)

4,300 miles (6,920 km) of this route, ancient peoples could trade for a variety of items, including silver from Spain, spices from India, and silk from China. Italian merchant Marco Polo traveled the Silk Road in the 13th century. One of the few Europeans to travel so far into Asia at the time, he later wrote about his extraordinary adventures in *The Travels of Marco Polo*. Many people who had no experience with the foreign cultures and varied lands encountered on the Silk Road thought he had concocted a work of fiction. Among other wonders, Polo described springs flowing with black oil—which are now known as the Baku oilfields—in Azerbaijan. "I have only told half of what I saw, because no one would have believed me," Polo reportedly said on his deathbed in 1324.

Archaeologists believe that the first wheel rolled out of Mesopotamia, the area that today comprises much of Iraq, as long ago as 3500 B.C. A picture on a clay tablet found in the ancient city of Uruk displays the profile of a sledge with two small wheels. Researchers also have found the remains of wheeled vehicles they believe date back to around 2700 B.C. The concept of the wheel spread to other civilizations, such as those in China and Egypt, where scientists have discovered evidence of early wheel use.

Despite the use of wheels and animals, the fastest way to send a message was often by foot. In 490 B.C., during the Greco–Persian Wars, a small Greek army defeated an invading Persian force at the city of Marathon. Legend states that the Greeks sent their best

The Silk Road (opposite) was made famous in part by Marco Polo, while Christopher Columbus (left) etched his name into the annals of history with his sea travels

runner to deliver news of the victory to the people of Athens, more than 25 miles (40 km) away. Although he had just returned from a 150-mile (241 km) trip, the fleet-footed messenger ran to Athens without stopping. After shouting, "Rejoice, we are victorious," he dropped dead. This story inspired the modern marathon, which is a 26.2-mile (42 km) road race.

Civilizations near seas adopted methods of water transportation. The Phoenicians, who lived on the Mediterranean Sea in what is now Syria and Lebanon, used fabric and linen to make sails for their ships starting in about 3000 B.C. By A.D. 850, the Vikings, warrior-sailors from Norway, Sweden, and Denmark, employed two different types of ships. Knorrs were wide but shallow vessels used for trading expeditions or when **migrating** to new lands. The faster longships, with room for up to 60 oarsmen on each side, were narrower and used for long-distance travel across the sea.

With the development of stronger and more durable ships, as well as improvements in **navigational** instruments and maps, Europeans ventured farther into unknown waters. The period from the 1400s to the 1600s is known as the Age of Exploration. It was during this time that Italian explorer Christopher Columbus encountered the "New World" in 1492 while attempting to find a western route to Asia and ending up in the Americas instead.

Later, people realized that they would save time if they did not have to sail around South America to travel from the Atlantic Ocean to

Construction of the Panama Canal was a grueling process, as workers labored to cut through Central American jungles and mountains using steam-powered machinery

the Pacific Ocean. Panama, a country in Central America, spans as little as 30 miles (48 km) at its narrowest point. People decided to build a **canal** through it to connect the two great oceans, saving 8,000 miles (12,875 km) and as much as 5 months of time at sea. After more than 30 years of work, the Panama Canal was completed in 1914.

Back on land, people were making new, innovative vehicles with wheels in the 19th century. The bicycle debuted in Paris, France, in 1817. Called a hobbyhorse, this early model did not have pedals, so riders had to push themselves along with their feet. It would be nearly 70 years before the machine would resemble the modern bicycle, with metal chains linking gears, pedals between the wheels, and rubber tires filled with air.

The early 1800s were marked by the appearance of the steam-powered train. The first trains, fueled by burning coal or wood that created steam, were huge, slow machines that traveled about four miles (6.4 km) per hour. But by the 1850s, the so-called "Iron Horse" was capable of attaining speeds of 50 miles (80 km) per hour or more. Train travel gained momentum, and by 1869, the United States had railroad tracks that ran all the way from New York City to Oakland, California. Likewise, the Trans-Siberian Railroad crossed Russia in northern Asia with 5,180 miles (8,290 km) of railroad track, from Moscow to Vladivostok, by 1900.

In the late 1800s, inventors developed the internal combustion engine, which ran on gasoline and was powered by small, contained

explosions inside a metal tube. The earliest cars with such engines were very expensive because each one was built by hand. American engineer Henry Ford introduced the mass-production system, in which many parts for vehicles were made and put together on an assembly line. In 1913, the Ford Model T became the first car to be manufactured in this way. Mass production allowed up to 1,000 cars to be completed each day, which led to cheaper vehicles that people with an average salary could afford.

While Ford was pioneering automobile manufacturing, American brothers Orville and Wilbur Wright gained fame by achieving the first

Transportational Headlines

THE COMPASS POINTS THE WAY

The compass was perfected in Italy, but it was originally developed in China. As early as the first century A.D., people realized that a metallic mineral called lodestone, which is naturally magnetic, pointed north and south. The earth is itself a giant magnet, and the magnet of the compass is attracted to the South Pole, making the compass point north. By 1300, the compass had made its debut as a navigational instrument. The compass gave sailors confidence in knowing which direction they were heading and the ability to find their way home more quickly, allowing them to sail farther into uncharted waters.

Four centuries after Columbus used a compass (opposite, bottom) in finding America, Henry Ford and his Model T (opposite, top) revolutionized everyday transportation

successful powered aircraft flight. The Wrights were bicycle makers and believed that if a rider could control a bike, then a pilot could control a plane. They built a plane with two sets of wings and tested it on December 17, 1903, at Kitty Hawk, North Carolina. The longest of their flights that day lasted 59 seconds. Sixteen years later, British pilots John Alcock and Arthur Brown flew their own plane all the way across the Atlantic Ocean without stopping.

Once airplanes became a reliable form of transportation, scientists set their eyes on a new travel destination: outer space. People in the 1960s witnessed a "space race" between the Soviet Union and the U.S., as each powerful nation vied to be the first to send vehicles and humans into outer space. In 1961, the Soviets sent a cosmonaut into orbit around Earth. In 1969, the U.S. landed two astronauts on the moon. People continued to make trips into space throughout the 20th century, but the expense of space programs and several tragic accidents led to a lull in human space travel shortly after the start of the new millennium.

Transportational Headlines

FIRST IN SPACE

Russian Yuri Gagarin made history on April 12, 1961, as the first human to travel into space. After studying engineering at college, Gagarin became a fighter pilot and later was selected for cosmonaut (the Russian term for astronaut) training. Just a year later, he was launched into space aboard the spacecraft **Vostok** *and spent 108 minutes in orbit around Earth. A problem with an electrical cable caused momentary spinning that could have ended in disaster, but the cable burned away as* **Vostok** *re-entered Earth's atmosphere, and the escape hatch successfully ejected, allowing for Gagarin's safe return.*

Fifty-eight years after the Wright brothers (pictured, left, in 1909) started the airplane age, Yuri Gagarin (below) proved that humans could indeed reach space

The Huntsville Times

CHICAGO DAILY NEWS SERVICE — HUNTSVILLE, ALABAMA, WEDNESDAY, APR. 12, 1961 — ASSOCIATED PRESS — WIREPHOTO

Man Enters Space

'So Close, Yet So Far,' Sighs Cape

U.S. Had Hoped For Own Launch

CAPE CANAVERAL, Fla. (AP) — The Redstone which the United States had hoped would boost first man into space stands on a launching pad. The Soviet Union beat its firing date by at least weeks.

"So close, yet so far," commented a technician who was helping groom the Redstone to send one of America's astronauts on a short sub-orbital flight, hopefully late this month or early in May.

"If we hadn't had those troubles last fall and on the chimp and Little Joe shots this year, we might have made it," the technician said.

"But you have to give the Russian scientists credit. They've accomplished a remarkable breakthrough."

Dr. Hugh Dryden, deputy director of the National Aeronautics and Space Administration, told the House Space Committee in Washington Tuesday that the earliest possible date for the manned launching is about April 28.

Project Mercury officials had hoped to achieve a manned Redstone flight last December or January. A series of launch mishaps necessitated additional launchings to qualify the system.

On Nov. 8, a space capsule failed to separate from a Little Joe rocket fired from Wallops Island, Va., in a test of the escape system.

Two weeks later, a Redstone fizzled because of a faulty connection which caused the escape

Hobbs Admits 1944 Slaying

By BOB WARD
Of The Times Staff

...authorities at Eglin Air Base, Fla. He signed a statement there detailing his slaying of the prominent 52-...old widow, Weaver said he...d from Air Force officials.

...suspect, now 43, is held by Air...authorities at Eglin Air Base, Fla. He signed a statement there detailing his slaying of the prominent 52-year-old widow, Margaret Thornton Fleming. Solicitor Macon L. Weaver...

...suspect, who has undergone psychiatric treatment by...ry authorities since Feb. 6,...covered his memory in full.

This is Russian Maj. Yuri Gagarin, history's first man in space. The Russians today rocketed him around the earth in an orbit taking slightly less than 90 minutes and brought him back safely to a prearranged spot in the Soviet Union. (AP Wirephoto via radio from Moscow)

Praise Is Heaped On Major Gagarin

First Man To Enter Space

'Worker'...

Soviet Officer Orbits Globe In 5-Ton Ship

Maximum Height Reached Reported As 188 Miles

MOSCOW (AP) — A Soviet astronaut has orbited the globe for more than an hour and returned safely to receive the plaudits of scientists and political leaders alike. Soviet announcement of the feat brought praise from President Kennedy and U. S. space experts left behind in the contest to put the first man into successful space flight.

By the Soviet account, Maj. Yuri Alekseyvich Gargarin, rode a five-ton spaceship once around the earth in an orbit taking an hour and 20 minutes. He was in the air a total of an hour and 48 minutes.

The whole sequence of events and the announcements relating to it raised a number of questions. The Soviet announcement said the flight took place today between 9:07 and 10:55 a.m., but some persons in Moscow's Western colony were skeptical that the feat actually came off today.

There was a curious sequence of events leading up to the announcement.

Rumors had been circulating several days that the space coup had been pulled off. Two days ago, Soviet TV technicians moved into the Central Telegraph Office with the evident purpose of getting pictures of correspondents in action as they reported such a story. There were various reports, none verifiable from official sources, that the flight had been made.

Then Tuesday night the Daily...

VON BRAUN'S REACTION:

To Keep Up, U.S.A. Must Run Like Hell

CHAPTER TWO

ENHANCING SAFETY AND EFFICIENCY

Following the explosion of transportational advancements during the early 20th century, many scientists began focusing not on revolutionary new modes of travel but on improving the systems already in place. People still use roads based on the relatively simple system begun in ancient Rome. However, engineers today use more durable materials and precise planning to build roads that help ease the flow of traffic. Today, the U.S. has almost 4 million miles (6.4 million km) of roadways. Many of these roads are part of the Interstate Highway System, the largest of its kind in the world.

Cars look much different today than they did 75 years ago, when the Chrysler 66 Coupe was considered a state-of-the-art model. Modern cars can go faster and feature safety equipment such as

Although the United States boasts the world's most extensive interstate highway system, countries such as China (opposite) have vast highway networks as well

Earth's supplies of coal (right) and other fossil fuels may be gone by as early as 2050, making electricity an increasingly attractive—and essential—form of fuel

seat belts and air bags. They are more comfortable, equipped with cushioned seats, heaters, and air conditioning. Today, many carmakers are working to build energy-efficient vehicles that use much less gasoline than cars of the past—or no gasoline at all. Why? Gasoline is made from petroleum, which is a **fossil fuel**. The earth contains a limited supply of oil, and when it is used up, it cannot be replenished. Experts predict that if people continue using oil at the current rate, the world could run out in about 40 years. Also, machines that burn fossil fuels give off toxic fumes that pollute the environment. Cars that use little or no fossil fuels are called "green" cars because they are better for the environment.

While the electric car is one of the most popular green car concepts, it is not new. In fact, electric cars were seen on roadways during the early 1900s. They ran at low speeds and were powered by a rechargeable battery that needed to be plugged into a charger. The Columbia Runabout, introduced in 1903 by an alliance of the Pope Manufacturing Company and the Electric Vehicle Company, could go 15 miles (24 km) per hour for 40 miles (64 km) before it needed a charge. The challenge facing today's engineers is figuring out how to make an electric vehicle that can go faster and farther without a recharge.

The Toyota Prius is an example of a **hybrid** car, which runs on both electricity and gasoline. At slow speeds, the car uses battery power. For more speed, the gas-powered engine gives the battery a boost. Aptera is a recently founded, California-based company that makes electric and hybrid cars. Not yet for sale in the mainstream market, the Aptera 2e model conserves energy, thanks to both its design and engine. It is a two-person car with three wheels instead of four. With its futuristic shape—resembling a sideways teardrop—the vehicle slides through the wind. Aptera's electric car can go about 100 miles (160 km) before the battery needs recharging. The hybrid version, which has a small gas **generator** to help keep the battery charged, has a range of 600 miles (960 km). The company planned to have both vehicles ready for sale in 2012.

The fuel cell is another new technology being used for greener vehicles. Instead of burning fuel as traditional cars do, fuel cells convert their fuel—usually **hydrogen**—to electricity, which powers the motor. The Honda FCX Clarity, which was available for lease only as of 2012, runs on a hydrogen fuel cell and can travel 74 miles (118 km) per gallon (3.8 l) of fuel. Its only **emissions** are water vapor and heat. However, the car is expensive; for now, the manufacturing of a single vehicle costs roughly $1 million. Besides the expense, these vehicles are currently limited by a lack of fueling locations. Before vehicles powered by this type of fuel cell can become widespread, a system of hydrogen refueling stations will need to be put in place.

Another mode of personal transportation under development is the pod car. With the capacity to hold six passengers, pod cars look like enclosed golf carts with no steering wheel. They run on an electric motor and glide along special guided tracks. Britain is currently testing a fleet of pod cars at Heathrow Airport near London. A passenger can simply get in the car, punch in the destination, and allow a computer to navigate the car along the path. Widespread use of pod cars could greatly reduce traffic **congestion**, pollution, and operating costs in crowded urban areas around the world.

Some countries are looking to high-speed trains to reduce automobile traffic and pollution. Japan has had bullet trains for years, and the citizens of France can zip along a railway on the Train a Grande Vitesse (TGV). The Japanese and French trains typically

For both speed and fuel efficiency, a streamlined form in vehicle design is vital, as illustrated by the Aptera 2e electric car (above) and this Japanese high-speed train (left)

More than 40 years after Neil Armstrong (below) became the first human to walk on the moon, Earth-bound modes of transportation are given more emphasis than space vehicles

zip along at up to 200 miles (320 km) per hour, making for quick travel between large cities. In 2009, China unveiled its own such train running from Guangzhou to Wuhan—a distance of roughly 664 miles (1,069 km). Traveling at speeds of up to 245 miles (394 km) per hour, the train became the fastest in the world, carrying passengers along the full length of the railway in just 3 hours. In 2009, U.S. president Barack Obama announced plans for the development of a high-speed rail system in America as well.

Flights into space do not make news headlines the way they did in the 1960s and '70s, but space travel and the study of outer space continue. Most recent human travel in space has been to the International Space Station (ISS). Although the ISS is the 10th space station to orbit Earth, it is the first **collaborative** effort among multiple countries. Construction began in 1998, and astronauts have been living and working there since the year 2000. It has taken more than 50 launches to deliver more than 100 parts that

COMPETITION FUELS IMPROVEMENT

Ever since ancient times, people have traveled to find food, escape enemies, and search for new homes. For centuries, most people stayed close to home, while only adventurous explorers traveled by ship to trade goods or overtake other cultures. And although humankind's competitive spirit has often led to war and other negative outcomes, it has also helped fuel developments in transportation. During World War I (1914–18) and World War II (1939–45), scientists vied to make dramatic upgrades to airplanes. During the 1950s and '60s, the world's most powerful countries—the U.S. and the Soviet Union—raced to be the first to launch rockets into space and send people to the moon, making great leaps forward in space travel.

From Amelia Earhart (below) to the multinational crews aboard the ISS (right), pioneers in the field of transportation build upon each other's work

were pieced together by astronauts on **spacewalks**. Crews from 16 nations have already spent time in the ISS, getting there aboard either a Russian *Soyuz* capsule or a U.S. space shuttle.

Trained astronauts are not the only people who have visited the ISS. In 2001, American millionaire Dennis Tito, who once worked as an aerospace engineer and later started an investing company, paid the Russian space program approximately $20 million to join the *Soyuz* for an 8-day trip to the space station. "For me, it was like being in heaven," Tito said of the experience. "It was like being in a second life." Since then, a few others have made the trip, including Iranian-born American businesswoman Anousheh Ansari.

In 2004, Ansari's family sponsored the Ansari X Prize, an award given to the first private organization to launch a manned craft into space and successfully return it to Earth twice within two weeks. SpaceShipOne, a spacecraft designed by American aerospace engineer Burt Rutan, won the prize in 2004, and Mike Melvill, the

THE RISKS OF DISCOVERY

*Whether sailing, flying, or ascending into space, pioneering travelers have long taken risks ... and sometimes paid for it. The **Titanic**—the largest ocean liner in history at the time of its completion—was declared unsinkable, yet it plunged to the bottom of the ocean after hitting an iceberg on its maiden voyage in 1912, killing 1,513 people. In 1937, American aviator Amelia Earhart disappeared over the Pacific Ocean while trying to fly around the globe. In 2003, the **Columbia** space shuttle broke apart on its return from a space mission, killing the seven astronauts aboard. Throughout history, such tragedies have invariably revealed flaws and led to better equipment and improved safety.*

Private developers have been responsible for some of the most significant recent advances in air and sea travel, creating SpaceShipOne (left) and submersibles (opposite)

craft's pilot, became the world's first **commercial** astronaut. While SpaceShipOne now hangs in the Smithsonian National Air and Space Museum in Washington, D.C., its successor, SpaceShipTwo, is in development. Perhaps within the next 10 years, as many as 6 passengers at a time will be able to reserve seats for a 62-mile-high (100 km) trip.

Some scientists, including Graham Hawkes, a British-born engineer and inventor, believe people should devote greater effort to exploring the oceans rather than space. Hawkes, head of the San Francisco-based Hawkes Ocean Technologies, has developed a variety of one- and two-person **submersibles** that can withstand the intense water pressure found in the ocean depths. The latest design is called *Deep Flight II*, and as of 2012, Hawkes was waiting to receive the funding necessary to build it. Hawkes has said that someday, people will live in the ocean rather than in outer space, and that countries aiming rockets into space are "180 degrees off course."

BETTER FUELS AND FARTHER FLIGHTS

Scientists are today looking to the future with an eye toward fuel-efficient vehicles, both on the ground and in the air. The Progressive Insurance Automotive X Prize competition in 2010 featured several strange-looking vehicles built for efficiency rather than style. The competition awarded money to participants who produced a vehicle capable of driving 100 miles (160 km) on a single gallon (3.8 l) of fuel. The vehicles could be powered by any kind of fuel as long as they did not sacrifice safety to gain efficiency. The winner of the mainstream category, the Very Light Car made by Edison2, a team of Americans with experience in professional sports car racing, looks like something from a science-fiction movie, with a body resembling an airplane and tiny wheels sticking out to the side. Another entry, the Wave II, appears to hover instead of drive on its three tiny wheels. The cars in the competition likely won't hit the market anytime soon, but they might someday provide the model for all vehicles on the road.

While some engineers have redesigned cars, others are attempting to redesign the fuel that cars use. Biofuels, which are made from plants, are not new; many countries have been using ethanol—made from corn—mixed with gasoline for years. However, scientists are researching the use of other types of biofuels, including cellulosic ethanol, which is made by breaking down the molecules of plant cells. If it works, this type of fuel could be made from

The futuristic Very Light Car (above) is an example of getting maximum value out of fuel, while biofuels (left) represent a means of replacing fossil fuels completely

FOOD VERSUS FUEL

To help reduce the use of fossil fuels such as oil in the transportation industry, researchers have looked to biofuels as a solution. Made from plant crops such as corn or sugar cane, biofuels are touted as being friendly to the environment as well as offering freedom from dependence on the world's dwindling supply of oil. However, as the demand for biofuels grows, poorer countries around the world will likely use more land to grow crops for fuel rather than food. This could potentially lead to food shortages in nations that already lack proper nutrition.

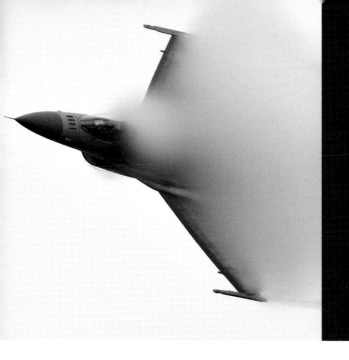

Planes, such as this F-16 fighter jet, that fly faster than the speed of sound actually have to fly through the sound waves they produce, which can make for a bumpy flight

the parts of plants—such as cornstalks, wood chips, and grasses—that are generally considered to have little use.

Airplane manufacturers are also considering ways to make planes that fly farther with less fuel and produce less pollution. As more people in **developing nations** fly on airplanes, the skies will become more crowded, and air traffic is expected to double by 2050. One idea is to develop planes that fly faster, which would deliver people to their destinations sooner. Airplane makers Lockheed Martin and Boeing have been investigating **supersonic** aircraft, which would be able to fly about 1.8 times the speed of sound—1,380 miles (2,208 km) per hour, or twice as fast as traditional planes. The problem with supersonic airplanes is the booming sound they create when they break the sound barrier. This sonic boom is illegal over populated areas of the continental U.S. Also, while this craft would save time, it would not necessarily save fuel.

America's National Aeronautics and Space Administration (NASA) has begun a study to develop planes that burn 70 percent less fuel and fly more quietly than today's aircraft. Possibilities include using **nuclear power** to run airplanes, using biofuels, or redesigning the bodies of planes to fly on less fuel. Airbus, an airplane manufacturer based in France, has designed the Airbus Concept Plane, projected for use by the year 2050. This plane has ultra-long and slim wings and a U-shaped tail, a form which could lower the rate of

fuel consumption, cut emissions, and decrease noise. One day, you might look up in the sky and see a "flock" of these airplanes flying in a V formation, just as geese fly to cut through the wind using less energy.

Another way for airplanes to use less fuel is to carry more people per flight. Twenty years from now, a long-distance flight might transport 1,000 passengers, with 40 people in each row. Because your chances of having a window seat would be slim, the plane might feature **virtual** windows to help keep passengers from feeling **claustrophobic**. It may be, too, that some future airplanes will not need any fuel at all. The Solar Impulse, currently under development by a group of private sponsors in Europe, is a plane made with superlight materials and powered by 12,000 solar cells.

Some engineers are attempting to combine cars and planes. An American company called Terrafugia has developed a "roadable aircraft"—that is, a car that can be flown or an airplane that can be driven. Approved by the Federal Aviation Administration (FAA) in 2010, Terrafugia's Transition drives on the road like a sport utility vehicle. Then, it can put out its wings and lift off into the air, fly at about 105 miles (168 km) per hour, and travel for roughly 500 miles (800 km) before it needs to land and head to the nearest gas station.

Often, technological advances in transportation originate with militaries. The Defense Advanced Research Projects Agency (DARPA), which provides research and development for the U.S.

The Terrafugia Transition (opposite) represents the first flying car, but for now, most people are content with car GPS devices (left) that can guide them to their destinations

Department of Defense, is working on a project that appears to be right out of the movies. For the Transformer project, DARPA hopes to combine the characteristics of a helicopter, airplane, and armored truck. The Transformer will carry four soldiers and feature rotor blades on top so that it can lift straight up into the air. The unit will be able to extend retractable wings, helping it to fly faster than a helicopter, and the controls in the vehicle will be simple enough that the driver will not need to go through extensive pilot training.

The development of more private companies—rather than governmental agencies—could lead to dramatic advances in space travel for ordinary citizens. Not too far into the future, people may

GLOBAL POSITIONING PROGRESS

Initially developed by the U.S. military, the Global Positioning System (GPS) was created in the 1970s and '80s. A satellite network called NAVSTAR allows people using special receivers to calculate their location with great precision anywhere on Earth. Using a process known as triangulation, the receiver calculates its distance from several satellites to determine its location, with accuracy ranging from one mile (1.6 km) to a few feet. In 1983, GPS became available for commercial use. Since then, the navigation system has been installed in cars, airplanes, and handheld devices.

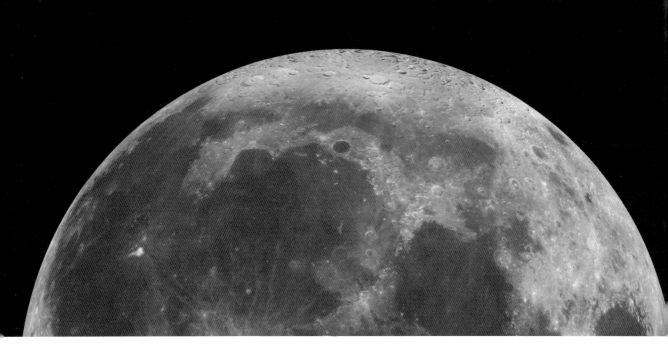

be able to pay for flights to tour the moon with a variety of companies that offer space experiences. These trips could provide a look at the moon from orbit, or they may land on bases established on the surface of the moon. NASA has announced plans to build a permanent base on the moon's south pole by 2024. "We're not going to be visiting space hotels on the moon for $10,000 in the next 10 years," said Eric Anderson, chief executive officer of the American space tourism company Space Adventures. "But we might be within the next 50 years."

While moon settlements are being constructed, astronauts may be exploring Mars. Of all the planets in our solar system, Mars is the most similar to Earth in terms of composition and distance from the sun. Mars, the moon, and even huge space rocks called asteroids have captured the attention of scientists because they could contain minerals and other resources useful for constructing bases and buildings in space. One proposed plan for traveling to Mars involves sending astronauts to a station orbiting Earth or possibly located on the moon, where another spacecraft would be waiting for them. Mars-bound spacecraft could be built in space, which would save the time and money necessary to launch them from Earth. An automated Earth Return Vehicle would be launched to Mars ahead of the astronauts. This unmanned unit would haul equipment and chemicals to create oxygen for the astronauts. Then, the astronauts would travel to Mars in a craft called a Mars

New space vehicles—and perhaps new fuels—will need to be developed before people can travel to the moon (opposite) or Mars (left) and remain there for any length of time

Habitation Unit, a trip that would take about six months when the planets are aligned in close proximity to each other. The astronauts would eventually come back in the Earth Return Vehicle, with the round trip taking between 18 and 24 months.

For even more distant flights, researchers are thinking up new designs for spacecraft. One is the sailcraft, or solar sail, which would need to carry no engine or fuel. Featuring a sail the size of a football field but as thin as plastic wrap, this vehicle would soar among the stars and planets powered only by the light in space. This craft likely would carry robots instead of people. If these flights were successful, it's possible that we could find other planets similar to our own. Then, the robots on board might terraform the planet—initiating a process of spreading specific bacteria and gases to change the environment of a planet to enable people to live there. Once such a planet is made ready, people could establish a colony in a truly new world. Such distant travel, however, is likely many decades in the future.

FROM HERE TO THERE

Some transportational developments, such as using cleaner-burning fuels and expanding the range of our space exploration, seem to be just over the horizon. But before these—and more fantastical advances—can become reality, society will need to embrace new technologies, infrastructures will need to be updated, and science will need to provide more answers.

Before cars that run on electricity or other types of clean energy become the new standard on roadways, people will need to buy them. Many people today buy vehicles based on their appearance, speed, and power rather than fuel efficiency and practicality. If people want to drive on long trips, they might not want battery-powered vehicles unless improvements lead to batteries that hold more energy and allow the car to drive farther without interruption. People might be reluctant to buy oddly shaped, futuristic-looking cars, no matter how little gas they burn. In addition, alternative energy vehicles are currently expensive to produce, which makes them unobtainable to buyers on tight budgets. If consumers do not demand these vehicles, then car companies might decide not to offer them. Carmakers developed models of electric cars in the 1960s and '70s, a time when the cost of fuel was rising. But consumers failed to embrace these vehicles, and gasoline-powered cars and trucks remained the norm.

Fuel cell engines produce more power than battery-powered engines, but they, too, are expensive to develop. Fuel cells, which usually run on hydrogen, produce only electricity and water, so they

Many people consider style as much as performance when deciding which car to buy; cars of the future will need to be not only efficient but attractive

do not pollute the air. However, the easiest method for making hydrogen fuel is from fossil fuels, which means that hydrogen fuel cells do not actually eliminate the need for oil.

Another major hurdle in the widespread adoption of new vehicles is development of the infrastructures they will require. If hydrogen fuel cell vehicles become popular, people will want convenient refueling stations. How many hydrogen stations will need to be built to provide easy access to car owners? And what about plug-in vehicles? Will gas stations change to meet the energy needs of every type of car, including fuel cell, plug-in, and biofuel? This type of undertaking is time-consuming and expensive. If pod cars become popular at airports, malls, and other crowded areas, these vehicles will need tracks to follow and computer systems to guide them. Super-sized airplanes capable of carrying 1,000 people at a time will require larger airports, longer runways, and perhaps more places for passengers to board and disembark.

Flying cars have been envisioned as the future of transportation ever since cars were first developed. But many obstacles stand in the way of widespread use of a flying car, or "roadable aircraft." Initially, these vehicles would be expensive for companies to manufacture (the Terrafugia Transition available today costs nearly $200,000), and only people with a pilot's license would be allowed to drive them. The price would go down as more people purchased them and learned to fly them, but city and state planners would then have to deal with the

Electricity began to be produced in power plants in the 1880s, about two decades after Jules Verne (left) published his novel that largely predicted the future

dangers of having many vehicles in the same airspace. The vehicles might need to be equipped with special sensors to alert drivers about their proximity to other cars. More infrastructure, in the form of special roads for takeoff and landing, might be needed as part of highways or in other locations. In bad weather, the vehicles might be prohibited from flying, which could require additional enforcement.

Even if flying cars are not widely used in our lifetimes, we may be able to tour the moon and visit Mars. Currently, speed is a major drawback to this kind of space travel. In 1969, when the U.S. space capsule used in the Apollo 10 mission returned from the moon, the craft reached speeds approaching 25,000 miles (40,000 km) per hour.

A GLIMPSE OF THE FUTURE

French author Jules Verne published a work of science fiction in 1865 called From the Earth to the Moon. *In the story, three men build a spacecraft, and then they load it into a cannon called a moon gun and fire themselves into space. More than 100 years later, researchers involved with the U.S. space program were astonished to discover how close some of Verne's descriptions came to reality. The writer's craft carried the same number of passengers, was built from the same material (aluminum), and was roughly the same size and weight as the spacecraft* Columbia *that carried the first astronauts to the moon in 1969.*

Rocket science is considered to be the realm of geniuses, and indeed, powerful and dependable rockets have been at the heart of successful space travel since the 1950s

Space shuttle speeds were about the same in the decades that followed. At that relatively snaillike rate, traveling to Pluto would take roughly 14 years.

To make spacecraft go faster, scientists will need to develop new propulsion systems capable of propelling a vehicle through space. One idea is to fuel ships with nuclear power. A series of small, controlled atomic explosions within such a ship's engine would significantly increase the speed of space travel. American physicists Theodore Taylor and Freeman Dyson were the first to suggest that this new power source be applied to spacecraft in the 1950s. However, using nuclear material was highly controversial at the time, and the idea failed to move forward because of the potential dangers involved. Now, researchers are taking a closer look at nuclear power for space transportation. A ship running on nuclear power could fly from Earth to Mars (a distance of about 34 million miles, or 55 million km) in just 3 months.

Another possible propulsion system is called the Variable Specific Impulse Magnetoplasma Rocket (VASIMR). With this system, electric power converts fuel into **plasma**, which is then ejected from the engine, thrusting the spacecraft forward. Hydrogen would likely be the fuel source for the VASIMR. Because hydrogen is plentiful throughout the solar system, a spacecraft could be refueled with it nearly anywhere it goes. NASA scientists are currently researching the VASIMR system, with testing perhaps soon to follow.

Saving time in traveling through space is important because this type of transportation can be harmful to the human body. In space, there is much less gravity than there is on Earth. Our bodies constantly work against Earth's gravity, which keeps our bones and muscles strong. But without the resistance of gravity, the body

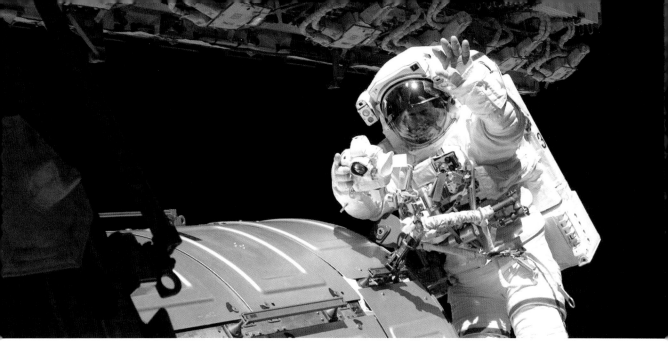

grows weak. Researchers are trying to find ways to create artificial gravity on spaceships to keep space travelers from becoming too weak by the end of their missions.

Another potential danger for space travelers is the **radiation** given off by the sun, other stars, and **black holes**. This radiation can cause cancer and other diseases in humans. Earth's atmosphere protects us from most of it, but in space, people would need much more protection. Researchers calculate that spending months in deep space could increase an astronaut's risk of developing cancers by up to 19 percent. "We can't yet estimate, reliably, what cosmic rays will do to us when we're exposed for so long," said Frank Cucinotta, a researcher with NASA's Space Radiation Health Project. Apollo

Transportational Headlines

MAGELLAN CIRCLES THE WORLD

In 1519, Ferdinand Magellan, a Portuguese explorer, took a crew of more than 200 men and 5 ships on a voyage around the globe, becoming the first European to sail from the Atlantic Ocean to the Pacific. To get to the Pacific, his ship had to sail around the southern tip of South America, a treacherous passage—where the two oceans collide—that is now known as the Strait of Magellan. After 3 years, 17 members of the expedition returned with 1 ship, the Victoria, completing the first circumnavigation of the globe and proving conclusively what many scholars had suspected: that Earth is indeed round.

The ISS (pictured, opposite, receiving repairs) is a relatively short jump from Earth, but the lessons learned there may be essential to future space travel

astronauts who went to the moon decades ago absorbed above-average radiation levels, but it's not clear if they have suffered long-term health problems as a result. However, traveling from Earth to the moon involved only a few days of exposure rather than the months that would be necessary for longer voyages.

Aluminum is a metal that is lightweight and strong, and engineers have used it for years to built spacecraft. Unfortunately, an aluminum ship might absorb only half of the radiation bombarding it and therefore fail to adequately shield its passengers. An expedition to Mars might require a different type of building material: plastic. Hydrogen is effective at absorbing cosmic rays, and plastics are rich in hydrogen. Plastic garbage bags contain a material called polyethylene, and this material absorbs 20 percent more cosmic rays than aluminum. Made into bricks, polyethylene would be stronger, lighter, and more protective than aluminum, and it could be the building material of the future in the field of space transportation.

Since ancient times, people have been motivated to visit distant lands and explore new areas. Fueled by the desire to gain knowledge, wealth, wares, and places to live, humans have steadily devised new and better forms of transportation. It is this history of constant innovation that permits people today to not only travel to all ends of the earth at speeds faster than sound but also to orbit our planet and the moon. What happens next—be it new vehicles, new fuels, or audacious ventures into space—is uncertain, but it will no doubt be exciting.

GLOSSARY

archaeologists — scientists who study material remains such as artifacts, monuments, and bones to learn about past human life and activities

black holes — former stars that have collapsed, forming regions of space whose gravity is so strong that nothing—not even light—can escape from it

cambered — having a slight arch on a surface

canal — an artificial waterway that is used for travel, shipping, or irrigation

claustrophobic — having an unusually strong fear of being confined in narrow or enclosed spaces

collaborative — describing a joint effort by two or more parties working together

colonies — groups of people who leave their native country and settle a new land but remain closely associated with or controlled by their homeland

commercial — relating to the buying and selling of goods and services, especially concerning private businesses

congestion — a state of being overfilled or overcrowded

developing nations — the poorest countries of the world, which are generally characterized by a lack of health care, nutrition, education, and industry

emissions — substances discharged into the air by the burning of fuels, as in a smokestack or automobile engine; emissions can contribute to air pollution

fossil fuel — a naturally occurring fuel, such as oil, coal, or natural gas, formed by the breakdown of ancient plants and animals; the gases from burning such fuels are believed to contribute to global warming

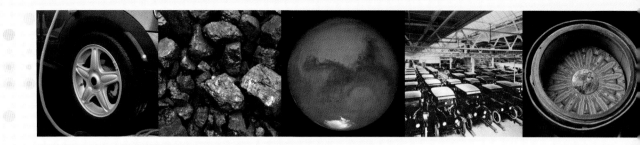

generator — a machine that changes mechanical energy, such as steady movement, into electrical energy

hybrid — something that has two different types of components that contribute in performing the same function

hydrogen — a colorless, highly flammable gas that is the most abundant element in the universe

infrastructure — the underlying foundation or basic framework of a system, such as that used for transportation or communication

migrating — moving from one country, place, or locality to another

navigational — describing a tool or technique that assists in determining position, course, or distance traveled for a ship, vehicle, or aircraft

nuclear power — energy produced by the splitting or fusing of atoms, the smallest particles that make up an object

plasma — a collection of charged particles that is a good conductor of electricity

propulsion — the action or process of driving forward

radiation — the process of sending out energy in the form of waves or particles; in excessive amounts, it can harm or kill living organisms

spacewalks — tasks or missions performed by astronauts outside their spacecraft while in space

submersibles — vessels or craft designed for underwater work or exploration

supersonic — describing speeds from 1 to 5 times greater than the speed of sound through air, which is about 768 miles (1,236 km) per hour

virtual — relating to something done outside the physical realm but meant to achieve the effects of something done in reality

SELECTED BIBLIOGRAPHY

Aczel, Amir D. *The Riddle of the Compass: The Invention that Changed the World*. New York: Harcourt, 2001.

Fridell, Ron. *Seven Wonders of Transportation*. Minneapolis: Twenty-First Century Books, 2010.

Oliver, Rachel. "Biofuels: 'Green Gold' or Problems Untold?" *CNN.com*, February 24, 2008. http://edition.cnn.com/2008/TECH/02/24/eco.biofuels.

Perraudin, Frances. "Space Tourism: Will It Be Worth the Money?" *Time*, October 31, 2010.

Solway, Andrew. *Designing Greener Vehicles and Buildings*. Chicago: Heinemann Library, 2009.

Sparrow, Giles. *SpaceFlight: The Complete Story from Sputnik to Shuttle—and Beyond*. New York: DK Publishing, 2007.

Woods, Michael, and Mary B. Woods. *Ancient Transportation: From Camels to Canals*. Minneapolis: Runestone Press, 2000.

WEB SITES

Future Cars
http://www.futurecars.com/
Browse this site to learn about biofuel, electric, fuel cell, and hybrid cars of today, as well as flying cars, hoverbikes, and other possible automobile designs of the future.

Virgin Galactic
http://www.virgingalactic.com/
Visit this site to learn about the world's largest private space exploration company and to get more information on the Burt Rutan-designed SpaceShipOne and SpaceShipTwo.

INDEX

Age of Exploration 9
airplanes 4, 14, 23, 25, 31–32, 33, 38
  fuel 31, 32
    biofuels 31
    nuclear power 31
    solar cells 32
  models 31–32
    Airbus Concept Plane 31–32
    Solar Impulse 32
  supersonic 31
Alcock, John 14
Ansari, Anousheh 25
Ansari X Prize 25
bicycles 4, 6, 11, 14
Boeing 31
Brown, Arthur 14
cars 4, 6, 11, 13, 16, 19–20, 28, 30, 31, 32, 33, 36, 38
  fuel 11, 19, 28, 30, 31, 36, 38
    biofuels 28, 30, 31, 38
    ethanol 28
    electricity 19, 20, 36, 38
    gasoline 11, 19, 20
    hydrogen cells 20, 36, 38
  futuristic designs 19–20, 28, 36
  hybrid 19, 20
  models 13, 16, 19–20, 28
    Aptera 2e 19–20
    Chrysler 66 Coupe 16
    Columbia Runabout 19
    Ford Model T 13
    Honda FCX Clarity 20
    Toyota Prius 19
    Very Light Car 28
    Wave II 28
Columbia 39
Columbus, Christopher 9
Earhart, Amelia 25

Edison2 28
flying cars 4, 32, 38–39
  Terrafugia Transition 32, 38
Ford, Henry 13
Gagarin, Yuri 14
Global Positioning System 33
Hawkes, Graham 26
Hawkes Ocean Technologies 26
highways 4, 6, 16, 39
horses 6
human runners 8–9
internal combustion engine 11, 13
International Space Station 23, 25
Lockheed Martin 31
Magellan, Ferdinand 42
Mars 4, 34–35, 39, 41, 43
Melvill, Mike 25–26
moon 4, 14, 23, 33, 34, 39, 43
  bases 34
  landing 14, 23, 39
NASA 31, 34, 42
navigational instruments 9, 13
  compass 13
Panama Canal 11
pod cars 20, 38
Polo, Marco 8
  The Travels of Marco Polo 8
Progressive Insurance Automotive X Prize 28
roads 6, 16, 39
robots 35
Rutan, Burt 25–26
sailcraft 35
ships 9, 11, 25, 42
  Titanic 25
  Victoria 42

Silk Road 6, 8
Soyuz capsules 25
Space Adventures 34
space effects on bodies 41–43
space shuttles 25, 41
  Columbia 25
space travel 14, 23, 33–35, 36, 39, 41–43
spacecraft 14, 25, 34, 35, 39, 41, 42, 43
  building materials 43
    aluminum 43
    polyethylene 43
  propulsion systems 41
    nuclear power 41
    VASIMR 41
  speeds 39, 41
SpaceShipOne 25–26
SpaceShipTwo 26
spacewalks 25
submersibles 26
  Deep Flight II 26
Sweet Track 6
terraforming 35
Tito, Dennis 25
trains 4, 11, 20, 23
  high-speed 4, 20, 23
  steam-powered 11
Transformer 33
Trans-Siberian Railroad 11
Verne, Jules 39
  From the Earth to the Moon 39
Vostok 14
wheel invention 8
Wright, Orville 13–14
Wright, Wilbur 13–14

Published by Creative Education
P.O. Box 227, Mankato, Minnesota 56002
Creative Education is an imprint of The Creative Company
www.thecreativecompany.us

Design and production by The Design Lab
Art direction by Rita Marshall
Printed in the United States of America

Photographs by Alamy (Everett Collection Inc., DIZ Muenchen GmbH/Sueddeutsche Zeitung Photo, Green Stock Media, MS Bretherton, Jeff Rotman), Corbis (Mike Blake/Reuters), Dreamstime (Allison Achauer, Alfiofer, Patrick Allen, Folco Banfi, Hxdbzxy, Jovan Jaric, Keng Po Leung, Georgios Kollidas, Dariusz Kopestynski, Lcro77, Brett Pelletier, Photosoup, Norman Pogson, Soleilc, Hannu Viitanen, Youths), Edison 2 (Brad Jaeger), Getty Images (Victor R. Boswell Jr./National Geographic, Hulton Archive, MPI, Steven Cole Smith/Orlando Sentinel/MCT), iStockphoto (David Shultz), Library of Congress, NASA (AEC, Goddard Space Flight Center, Scientific Visualization Studio, USAF Photo), Shutterstock (Alperium, Kobby Dagan, Dongliu, rook76), U.S. Air Force (Joe Oliva)

Copyright © 2013 Creative Education
International copyright reserved in all countries. No part of this book may be reproduced in any form without written permission from the publisher.

Library of Congress Cataloging-in-Publication Data

Dittmer, Lori.
The future of transportation / by Lori Dittmer.
p. cm. — (What's next?)
Includes bibliographical references and index.
Summary: A look at potential future developments in transportation, including privately operated spacecraft, as well as electric cars and other technologies that are currently considered state-of-the-art.
ISBN 978-1-60818-224-4
1. Motor vehicles—Juvenile literature. 2. Transportation—Forecasting—Juvenile literature. 3. Transportation—Technological innovations. I. Title.

TL147.D57 2012
629.04—dc23   2011040509

First edition

9 8 7 6 5 4 3 2 1

Cover: The busy streets of Shanghai, China
Page 1: A Shanghai bicyclist hauling foam
Page 2: A motion-blurred photo of a high-speed train